## The Young Geographer Investigates

# Grasslands

## Terry Jennings

Oxford University P

**Oxford University Press, Walton Street, Oxford OX2 6DP**

Oxford  New York  Toronto
Delhi  Bombay  Calcutta  Madras  Karachi
Petaling Jaya  Singapore  Hong Kong  Tokyo
Nairobi  Dar es Salaam  Cape Town
Melbourne · Auckland

and associated companies in
Berlin  Ibadan

*Oxford* is a trade mark of Oxford University Press

ISBN 0 19 917083 5 (Paperback)
ISBN 0 19 917089 4 (Hardback)

© Terry Jennings 1988

First published 1988
Reprinted 1990

Typeset in Great Britain by
Tradespools Ltd., Frome, Somerset
Printed in Hong Kong

**Acknowledgements**

*The publisher would like to thank the following for permission to reproduce photographs:*

Ardea p.7 top, p.11 top left, p.26 bottom, p.38 top and right; Aspect Picture Library p.1 bottom, p.18 left, p.27 bottom, p.28 top, p.44 inset; BBC Hulton Picture Library p.34; BBC Enterprises © 1978 p.8; Biofotos p.4 centre and centre top, p.11 centre, p.13 left, p.16 top; Bruce Coleman p.16 bottom left, p.26 top, p.30 top and bottom; Douglas Dickins p.41 bottom; Robert Estall p.33 centre; Susan Criggs COVER, p.5 left, p.37 top left; Robert Harding, contents page, p.32, p.33 bottom, p.36 top; Hutchison Library p.16 bottom right, p.29 bottom right, p.37 bottom left, p.39 main picture; Impact Photos p.6; Jacana p.38 left; Terry Jennings p.5 right, p.10, p.13 right, p.17 bottom, p.18 top right; Frank Lane Picture Agency p.4 top, p.11 bottom left, p.27 top, p.28 bottom, p.39 inset; Tony Morrison p.15, p.17 top, p.35 top and bottom, p.36 bottom; Natural Science Photos p.11 top and bottom right, p.17 top and centre, p.18 bottom right, p.28 inset, p.37 right; Oxford Scientific Films p.29 top right, p.35 centre; Photo Library of Australia p.29; G.R. Roberts p.31 left and right, p.33 top; Arthur Shepherd p. 9; Spectrum Colour Library p.12 bottom left; Suttons Seeds Ltd p.41 top; Topham Picture Library p.12 top left, p.14; Wimpey Group services p.7 bottom.

Illustrations are by Richard Hook, Peter Joyce, Ben Manchipp, Ed McLachlan, Techniques, Cathy Wood.

# Contents

# Grasses

*Above*: Meadow grasses and buttercups, and (*right*) mat grass growing among rocks

Grasses are some of the commonest plants. There are about 10,000 different kinds of grasses. They cover more than one quarter of the land on which plants can grow.

Grasses need far less water than trees. Therefore grasses can grow where trees cannot. Grasses can also survive burning and freezing. They grow from the Arctic tundra to the tropics, from the edge of the sea to the tops of mountains. Grasses are found on the open plains and in the depths of the forest. They grow in wet places and dry places. The open sea is the only place on Earth where grasses have not been able to grow.

Grasses are the most important plants in the world. They provide food for humans and other animals. Indirectly they provide us with some of our clothes. They also help to stop the soil being washed away by heavy rains.

*Above*: Grasses growing in the mud of an estuary and (*below*) tussocks of grass on the Arctic tundra

4

# Why are grasses so successful?

A large part of each grass plant is underground. Grasses have masses of fibrous roots. In dry weather grasses stop growing. They also stop growing in winter. With warmer or wetter weather, the grass starts to grow again. If grass is burned, its underground buds can soon start to grow again.

Grasses have another big advantage over trees. Grass leaves grow from their bases near the ground. If a cow or another grazing animal nibbles the top of the grass plant, it grows up again from the bottom. The same thing happens if people cut a lawn or playing field with a mower. Other plants would soon die if they received this treatment. But grasses go on growing and growing.

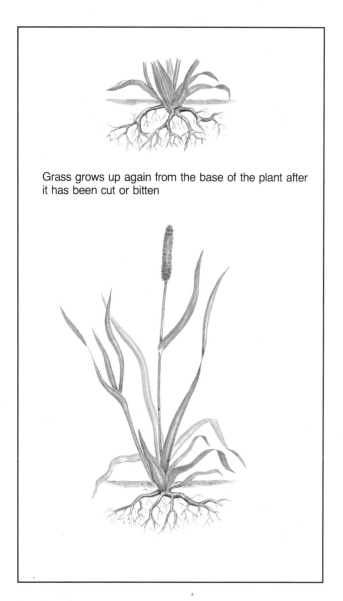

Grass grows up again from the base of the plant after it has been cut or bitten

Mowing a grass lawn

Cows grazing on grass

# Where grasslands occur

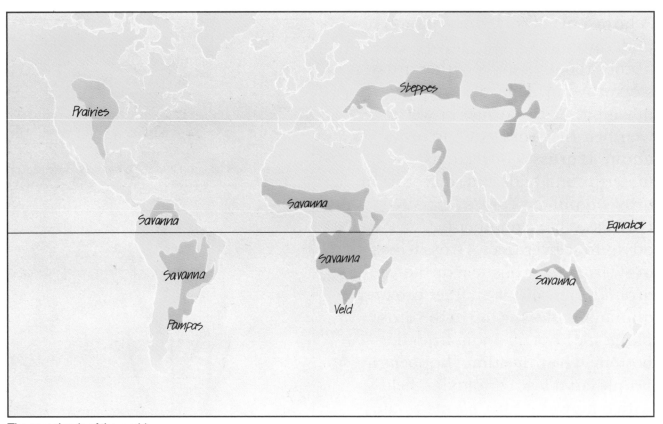

The grasslands of the world

Grasses need quite a lot of light if they are to grow, although certain kinds can grow in shallow water. They grow only where light can reach them. Grasses are found growing in estuaries, freshwater lakes and marshes. There are even some which grow at the edge of the sea.

But at the heart of all the continents, except Antarctica, there are huge areas of grassland. They include the prairies, the savannas, the pampas, the steppes, the veld and the Australian outback. The map shows where in the world these vast grasslands occur.

Grasses also grow on man-made pastures and lawns. These are cultivated grasses. So also are the cereals from which much of our food comes. Some grasses, growing where they are not wanted, are weeds of garden and farmland. Because they grow fast and produce many seeds, weed grasses can be difficult plants to get rid of.

Grass often grows as a weed

# Grasses and soil erosion

Grasses are important in another way. Growing almost everywhere, grasses stop the soil from being washed away when it rains on hillsides and steep slopes. This washing away of the soil is called soil erosion. Soil erosion can also occur when the wind blows the soil away.

Soil erosion caused by walkers

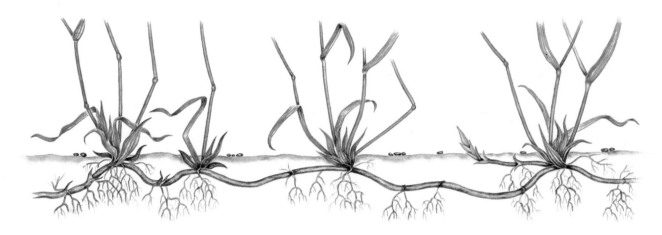

Many grasses spread by putting out stems along the ground (stolons) and under the ground (rhizomes)

Grass seeds are very light and easily blown about by the wind. Any bare patch of soil soon has grass seeds on it. The grasses grow quickly and cover the ground. As well as producing seeds, many grasses spread by sending out underground stems. These creep along under the soil and send up new stems. A single grass plant may soon form a whole patch of grass. It is this network of stems and roots which keep the soil in place when it rains or the wind blows.

The banks of new road and railway cuttings are sown with grass seed. The grass stops the soil from being washed away on to the road or

This new road embankment has been sown with grass

railway. Anything which destroys the cover of grasses over the soil will cause soil erosion. Too many animals grazing, or too many people or animals trampling over the grass can cause soil erosion. So can ploughing slopes and hillsides.

# Grasses in history

Our early ancestors had no settled home. They lived a roaming or nomadic life. The men hunted wild animals for meat and skins. The women and children searched for leaves, fruits, seeds and berries to eat. Life became easier when people learned how to keep sheep, goats and other animals.

Some of the grasses of that time had large seeds which could be eaten. Often the grass seeds were toasted on hot stones, like popcorn. About 10,000 years ago, someone noticed that new plants would grow if grass seeds were dropped onto the soil. Soon it was discovered that seeds could be stored. Now the seeds could be eaten through the long winter months when little other food was available.

As people learned how to plant and harvest the grasses and other wild

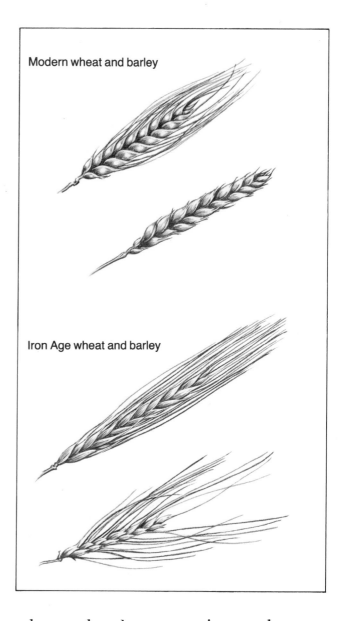

Modern wheat and barley

Iron Age wheat and barley

Our Iron Age ancestors may first have grown cereals in forest clearings like this

plants, they began to give up the wandering life. They started to build houses to live in. They cleared parts of the forest to plant their crops. This was the beginning of farming. The early farmers discovered that they could get better crops if they saved the seeds from the biggest and best plants. Gradually, by picking out the best, our ancestors were able to improve the quality of their crop plants. This process is still going on today, helped by scientists.

# Improving grasslands

Spraying grassland with weedkiller

Cutting grass to make hay

Baling hay

Piling up silage

Grass is a very important food for cattle and sheep. These animals supply much of our meat, milk, butter and cheese. They also supply some of our materials such as wool and leather.

It has been found that some grasses are better for feeding farm animals than others. Many of these grasses have been specially bred by scientists. Often the farmer sows a mixture of seeds. There may be two or three kinds of grass seeds and also clover seed. These grow into the kind of pasture that farm animals like. The farmer puts fertilizer on the grass to make it grow better. Chemical sprays may be used to kill off the weeds. In dry weather, the grassland may be watered to make sure it keeps growing.

On hills and mountain slopes, where the soil cannot be ploughed, the farmer may improve the natural grassland. He may use weedkillers to kill off large-leaved weeds. He spreads fertilizer on the grass so that it grows better. Many of the world's natural grasslands have now been improved in this way.

During the summer the grass may be cut to make hay or silage. Hay is dried grass. Silage is fresh grass which is stored where the air cannot get to it. Hay and silage are used to feed farm animals in winter when the grass stops growing.

# Reed swamp

Reeds are found in most parts of the world. They grow along the edges of lakes, dykes and ditches. They also grow in marshes, swamps and shallow water. Often reeds are planted on the banks of rivers. They spread quickly by their fast-growing underground stems. The stems and roots of the reeds bind the mud together and stop it being washed away.

As well as being beautiful, reeds are also useful. In many parts of the world reeds are harvested. The dead stems of the reeds are used for thatching house roofs. Mats, baskets and certain kinds of fences and furniture can also be made from them.

The feathery flowers of reeds make good decorations in houses during the winter months. The stems of papyrus, or paper reed, were first used thousands of years ago to make paper. In addition, reeds provide food and shelter for many marshland birds and animals.

Reeds and kingcups growing on Suffolk marshland

A stack of cut reeds drying in the sun

Reeds are clamped to the roof with sharp wooden staples

A thatcher at work and (*inset*) a reed cutter

# Bamboo

Shrimp basket in Madagascar

Long house in Sarawak

Bamboo growing in Bali

Scaffolding in Shanghai

Making umbrellas in Thailand

Bamboos are large woody grasses. Some kinds grow to be 36 metres tall. Like many other grasses, bamboo shoots grow from stems which creep along under the ground. All bamboos have smooth, hollow jointed stems.

There are about 500 kinds of bamboo. They grow mostly in or near the tropics. In China, Japan, India, Sri Lanka and parts of South America there are large bamboo forests. Some bamboos, from cooler climates, are grown in gardens in Britain, France and other parts of Europe.

Many bamboos flower every year, producing grains that can be eaten.

Often in times of drought, when other crops have failed, bamboo grains have saved the people of Asia from starvation. The Chinese eat the tender young shoots of bamboo as a vegetable.

But bamboo has many other uses. Houses can be built from bamboo, while paper can be made from the stems. The split canes are used to make mats. Pieces of the large stems are used as buckets and jars. Bamboo stems are also made into knives, blow-pipes, fish traps, baskets and ropes. In addition, bamboo can be used to make fishing rods, boats, water pipes and musical instruments.

# Sugar cane

Sugar is made from sugar cane, a giant grass. Some kinds grow to be over 6 metres tall. The stems are jointed and brightly coloured. Yellow, green, red and purple are the commonest colours. The stems may be 5 centimetres in diameter at their bases. They contain a sweet, sticky juice.

Colourful stems of sugar cane

Machine harvesting

Sugar cane is grown in many parts of the tropics. It thrives in rich, moist soil and sunny conditions. It is grown on large farms called plantations. There are sugar plantations in the West Indies, the Southern United States, Hawaii, Jamaica, Cuba, Brazil and parts of Australia.

The sugar cane is cut by hand in many places. But machines are being used more and more. The cut cane is quickly taken to the mills. At the mill, the cane is cut into short pieces.

Cutting sugar cane in Costa Rica

Fire on a sugar plantation in Barbados

These are crushed by huge rollers to squeeze out the juice. Finally the juice is boiled until it forms crystals of sugar.

# Cereals

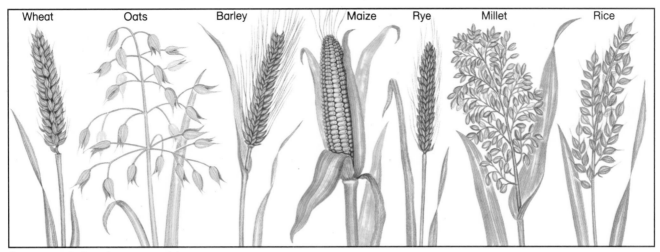

Some cereals (not drawn to scale)

Wheat, oats and barley are all kinds of grasses. These food grasses are called cereals. Rye, maize, millet and rice are also cereals. The seeds of these different cereals provide food for people all over the world.

The seeds of wheat, or wheat grains, are ground into flour. From flour we make bread, cakes, macaroni and spaghetti. Rice is the main food for about half of the people in the world. Oat grains are ground to make oatmeal. We eat porridge made of oatmeal. From barley, malt for making beer is obtained. Cornflakes are made from the grains of maize. So is corn oil which is used for frying.

Rye can be used to make bread and also whisky, vodka and gin. Millet is a group of food grains which grow in the tropics. They grow in poor soils on which most other crops will not grow. In addition, cereals of all kinds are fed to farm animals. From these farm animals we obtain meat, milk, eggs, cheese and some of our clothing materials.

Field of millet

Hens and chicks feeding on cereal grains

# Wheat

Wheat is one of our most important cereals. Its grains are the ones most often used for making flour. From flour, bread, cakes and biscuits are made. Wheat is also eaten in other ways including spaghetti, semolina, macaroni and some breakfast cereals.

Wheat is now grown over much of the world except for the moist tropics and the polar regions. It grows best in those places where the planting season (autumn or spring) is cool and moist, and the summers are sunny and fairly dry. Many countries grow wheat for their own use. They include Britain, China, France, Russia and India. Some countries grow wheat and have a surplus for export. Among these are the United States, Canada, Australia, Argentina and the countries of the European Economic Community.

On a modern farm, many machines are used in the planting and growing of wheat. Chemical sprays and fertilizers are used to help the wheat to grow better. Scientists have produced many new kinds of wheat. The early wheat plants had long thin stems and few grains in each 'ear'. Modern wheat plants have short, sturdy stems which do not bend or break in the wind or rain. They have many more grains in each 'ear'. Some of these new kinds of wheat are less likely to catch the diseases which sometimes killed the wheat crops of the past.

A wheatfield and (inset) foods made from wheat

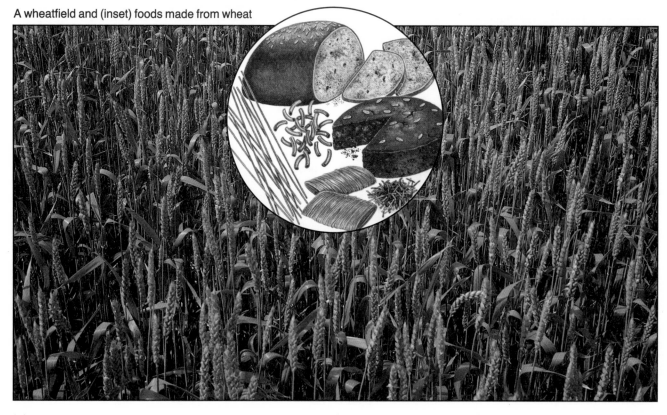

14

# Maize

Maize being harvested, and (*inset*) some things made from maize

Maize is another important cereal crop. In parts of eastern Europe, Africa, Asia and South America, maize is one of the main foods. The South Africans call maize 'mealies'. In America, maize is called 'corn'. In fact, the first people to grow maize were the American Indians.

Maize plants grow very tall. They may be 4 metres or more high. The large yellow grains form on a swollen stem called the cob. Maize grains are used to make cornflakes, popcorn and the corn starch from which custards and blancmanges are made. Certain chemicals and glues are made from maize. And maize grains are crushed to make the 'corn oil' used in cooking. Some kinds of maize have grains that taste sweet. These are known as sweet corn. They are cooked and eaten as a vegetable.

About half of the world's maize is grown in a part of the United States known as the Corn Belt. There the nights are warm and there is plenty of sunshine and summer rains. Most of this maize is fed to farm animals. The stalks and leaves of the plants are also fed to farm animals as silage.

15

# Rice

Rice is the main food for millions of people who live in China, Japan, India and South-East Asia. There are two main kinds of rice. One kind grows on dry, sloping fields. This is called upland rice or hill rice. However, most rice is grown on fertile soils by rivers and lakes. This lowland rice needs hot weather and a great deal of water.

In the Southern United States, rice is sown, cultivated and harvested by machines. But in most parts of the world, everything is done by hand. First the rice grains are soaked in water. They are then sown in special seed-beds of mud. Meanwhile the rice fields are flooded by rain or water from rivers or streams. The rice seedlings are transplanted to the flooded fields. When the rice is ripe, the fields are drained, and harvesting begins.

Harvesting is usually done by cutting off the grain with sickles.

Planting out rice seedlings in a flooded field

Later the straw is eaten by water buffaloes or other farm animals. When harvested, the rice is enclosed in a hard brown husk and is known as 'paddy'. That is why flooded rice fields are called paddy fields.

Two thousand year old rice terraces in the Philippines

Harvesting rice by hand in China

# Grassland animals

The world's largest plant-eating animals live on grassland. On the grassy plains there are few places where the animals can hide from their enemies. Elephants and rhinoceroses are too big to be attacked by wild animals. Most of the other large grass-eaters, such as antelopes, zebras and buffaloes can run fast. Speed is their main hope of survival. These grass-eaters have good eyesight which helps them to spot their enemies. Many of them live in herds. This means that there are always a lot of eyes and ears to be alert for danger.

Grass-eating animals have special teeth to deal with their food. The teeth have ridged surfaces. They are able to grind the grass into small pieces.

Three of the world's largest birds feed on grassland areas. All three kinds are unable to fly. Ostriches feed on the savanna grasslands of Africa, while the rhea lives on the South American pampas. The emu has its home on the dry plains of Australia. All three have good eyesight. And they are able to run fast to escape their enemies.

African buffaloes and their calves

Sheep have special teeth for grinding up the grass

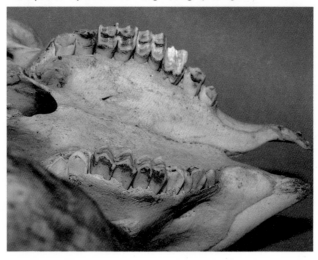

17

# Hunters of the grasslands

These cheetahs have just caught an antelope to eat

Two lions hiding in long grass

Vultures feeding on a dead zebra

Animals which eat grass and other plants are called herbivores. They in turn are food for many different kinds of flesh-eating animals, or carnivores.

The world's swiftest hunters live on grassland. On the African savanna, the herds of antelopes, zebra and gazelles are food for swift-running hunters. Cheetahs feed almost entirely on the smaller antelopes and gazelles. These carnivores usually hunt in twos or small family groups. A cheetah can run at 100 kilometres per hour. It can overtake a gazelle or antelope and kill it after a short burst of speed.

Lions spend most of their life sleeping or resting. They live in groups of up to 30 animals. It is mainly the females which hunt. They spread out and stalk their prey. Often they surround their prey. Then the males and cubs share in the meat.

Hyenas hunt, but they also scavenge. This means that they feed on dead animals and the leftovers of lions and cheetahs. Vultures and jackals also live by scavenging. In spite of its long legs and streamlined body, the maned wolf feeds largely on small creatures and fruit.

# Do you remember?

1 How many kinds of grasses are there in the world?

2 Where are the only two regions on Earth where grasses have not been able to grow?

3 What are the roots of grass plants like?

4 What happens if the top is cut off a grass plant?

5 Name three of the wet places where grass can grow.

6 Whereabouts on the continents are there large areas of grassland?

7 What causes soil erosion?

8 Why are the banks of new road and railway cuttings sown with grass seed?

9 How did the early farmers improve their crop plants?

10 What does a farmer spread on his grassland or pasture to make it grow better?

11 Name five things which we obtain from animals which eat grass.

12 What is the difference between hay and silage?

13 What are hay and silage used for?

14 What are the dead stems of reeds used for?

15 Name two of the countries where large bamboo forests occur.

16 Name two uses of bamboo.

17 How is sugar obtained from sugar cane?

18 What are cereals?

19 Name five kinds of cereals.

20 From which cereal is most of our flour made?

21 In what kind of climate does wheat grow best?

22 What do the Americans call maize?

23 Name four things which maize can be used for.

24 What are the two main kinds of rice?

25 What are the fields like in which most rice is planted?

26 What is "paddy" another name for?

27 What are the teeth of grass-eating animals like?

28 Name three large flightless birds which feed on grasslands.

29 What is another name for an animal which eats grass and other plants?

30 What is another name for an animal which eats the flesh of other animals?

# Things to do

Veins on a blade of grass, and a tree leaf (magnified)

**1  Grass leaves**  Collect some grass or cereal leaves. Also collect some leaves from trees, weeds and garden plants. Look at the leaves carefully. Look at the pattern made by the veins on the leaves. You may need to use a hand lens or magnifying glass for this. How are the leaves of grasses and cereals similar to those of other plants? How are they different?

Can you find any garden plants which have leaves similar in form to those of grasses and cereals?

**2  Grass flowers**  See how many different kinds of grasses you can find. This is best done when the grasses are in flower, mainly in June and July. Look at the different flowers carefully with a hand lens or magnifying glass. How do grass flowers differ from other garden flowers?

Draw some of the grass flowers or press them between sheets of newspaper or blotting paper covered by heavy weights.

After a few days, when the flowers are completely flat and dry, mount them on a sheet of card or in a scrapbook.

**3  Does grass grow faster when it is cut?**  In summer, on an out-of-the-way part of the lawn or playing field, mark out two identical squares of grass. Squares 30 by 30 centimetres are good ones to choose.

Leave one square of grass untouched for, say, eight weeks. At the end of the eight weeks, cut the grass down short with a pair of scissors. Weigh all your grass cuttings. How much do they weigh?

Cut the grass in the other square every week. Weigh each week's grass cuttings. Do this for eight weeks and record the total. Is it more or less than the weight for the other square over the same period of time?

Does grass grow faster or slower when it is cut? What effect do cows and lawn mowers have on the grass?

As well as eating the grass, what do cows do which makes grass grow faster? What do cows do which makes grass grow slower?

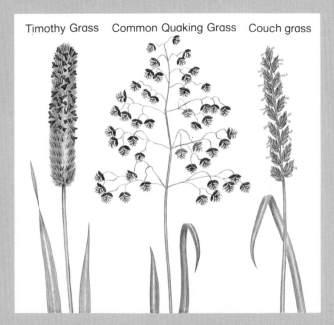

Timothy Grass   Common Quaking Grass   Couch grass

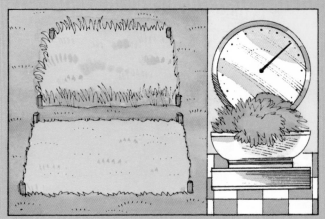

**4 Growing seeds** Plant a few seeds of different kinds in pots of moist compost. Plant a few grass seeds and seeds of cereals such as wheat, oats, barley and maize. In some more pots, plant seeds such as mustard, cress, radish, peas and beans.

Stand the pots on a sunny windowsill. Do not let the compost dry out.

When the seeds grow, look carefully at the seedlings. How do the seedlings of grasses and cereals differ from those of other plants? In what ways are they alike?

**5 Soil erosion** Find two identical seed trays. Fill one with garden soil. Flatten the soil down level with the top of the seed tray. Put garden soil in the other seed tray. In this soil, plant tussocks of grass.

Stand both seed trays out in the open garden or playground. If the weather is dry, water *both* trays.

After a week or two, when the grass in the seed tray is growing well, stand both trays up on bricks or blocks of wood as shown in the picture. Water both trays hard from a watering can or hose. From which tray does the soil wash away most easily? Can you now see why grass protects the soil on hills, slopes and mountains?

Which tray loses soil most easily?

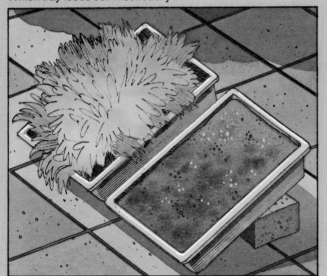

**6 Making a fire drill** Early people used a fire drill to make fire. Try to make a fire drill like that in the picture below.

When you use it you may be able to make the point of the drill hot. You may be able to make smoke. But you are unlikely to find the right kind of wood to make fire. Just to be on the safe side, though, only use your fire drill out on the lawn, playing field or playground.

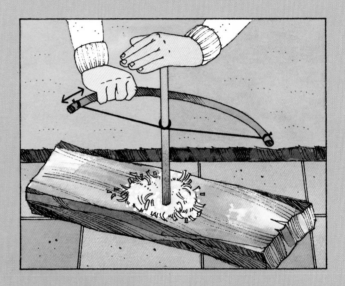

**7 A model thatched house** Make the walls of a model house with cardboard, clay or plasticine. Use dried grass, wheat stems or drinking straws to thatch your house. Paint your model house.

21

**8 Make a collection** Collect samples of as many kinds of cereal seeds as you can. Look for examples of wheat, oats, barley, maize, rye, rice and millet seeds. Mount and name them. Collect as many of the cereal plants as you can. Press these between sheets of newspaper or blotting paper covered by heavy weights. Mount and name the cereal plants as well.

Collect examples of wild grasses. Mount and name them. Compare these grasses with your cereals.

**9 Cereal seeds** Soak some wheat and maize seeds in cold water for a few hours until they have become swollen with the water they have taken up. Cut some of the seeds in half. Compare the insides of your seeds with the pictures below. Can you see all of the parts in your seeds that are labelled in these pictures?

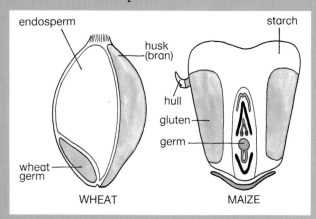

**10 Growing cereals** Try growing different cereal seeds such as wheat, oats, barley and maize. Grow them in clean plant pots or yoghurt pots. Use a seed-growing compost or a good garden soil and fill each pot to within about a centimetre of the top.

Soak the seeds for about 24 hours in clean, cold water. Sow four or five seeds of each kind in each pot. Press the seeds gently into the compost or soil, and cover them lightly with soil or compost. Place the pots on a sunny windowsill. Water the seeds each day.

How many of your seeds grow? Measure the seedlings every day. Show how your seedlings grow taller as the days go by, by plotting a graph of your results.

Do different cereals grow at different rates?

What happens if you put some of the seeds in pots in a dark cupboard?

If you carefully transplant your cereal seedlings to larger pots of moist soil or compost, they should go on growing. With luck you may get some ears of wheat and other cereals being formed in a few months time. How many grains are there in each of the ears?

Making a graph to show the growth rate of cereal seeds

**11 The cultivation of wheat** Draw and paint a number of pictures to show the different stages in sowing, cultivating and harvesting wheat. Make a display of your pictures around your classroom.

**12 Flour** Collect small samples from as many different kinds of flour as you can obtain. Examine them closely with a magnifying glass or microscope. What differences other than colour can you see?

**13 Making your own flour** Find a large flat stone which has at least one smooth surface. A piece of concrete paving slab will do if you cannot find a natural stone. Find another, smaller stone which you can hold comfortably in your hand. A rounded beach pebble is ideal.

Put a few grains of wheat on the flat stone and crush them with the smaller stone. You will find this is hard work. If the grain is very hard and dry it helps to soak it in water for a minute or two, and then to strain it and blot it before you continue milling. Soon the grains will open to reveal their white floury insides (the endosperm). This is the way our distant ancestors made their flour.

Making flour

If you want to make bread from your flour, then you must make absolutely certain that your two stones have been scrubbed clean before use.
NOTE: Do not make your flour from wheat which has been dressed with chemicals. If in doubt, buy your wheat seeds from a health food store.

**14 Making unleavened bread** Take about 50 grammes of plain flour in a cup or small basin. Add water, a little at a time, and mix it thoroughly to make a stiff paste or dough. Roll the dough into a ball on a floured working surface. Flatten the ball slightly and place it in a greased baking tin.

Ask a grown-up to help you to bake your dough. Bake it in a hot oven (230°C, 450°F, Gas Mark 8) for 10 or 15 minutes or until it is golden brown.

When it is cool, taste your bread. Bread like this is called unleavened bread. It must taste something like the bread people had to eat thousands of years ago before it was discovered how to make bread rise with yeast.

Try also baking dough made from 50 grammes of wholemeal flour or 50 grammes of oatmeal. What does this bread taste like?

**15  The world our cereal bowl**  Many of the foods we buy in the shops are made from cereal grains or, more rarely, from other grasses. Often the labels on the packets, jars or bottles tell us what the food is made of and where the food was grown or packaged.

Obtain a small map of the world and stick it in the centre of a large sheet of paper or card. Fasten the card or paper onto a notice board or a large piece of wood.

Around the edges of the sheet of paper or card, stick the labels or packet fronts of cereal foods that were grown or made in other countries. For each label put a pin in the country it came from. Then join the label to the pin by a length of thread. Sellotape the end of the thread on to the label.

# Things to find out

1 Think about the different meals you eat. Are there any of your meals which do not include cereals in one form or another?

2 The word 'cereal' comes from the name of a Roman goddess, Ceres. Find out who Ceres was.

3 Some kinds of breakfast cereal contain bran. What is bran? Why is it said to be good for us?

4 Different kinds of flour, including brown, wholemeal, white, plain and self-raising flour can be made from cereals. What is the difference between these different kinds of flour? What are they used for?

5 There is an expression which says that 'All flesh is grass'. Find out what this expression means. Do you think it is true?

6 When does a cow first give milk? How is a cow made to go on giving milk? What are calves fed on if they do not receive their mother's milk? Roughly how much milk does a cow make each day?

7 Often sand dunes by the sea become colonized by grasses. The dunes may gradually turn into grassland or even woodland. Find out about these grasses which colonize sand dunes and how they are able to grow in the loose sand at the edge of the sea.

8 Find out exactly how a roof is thatched. What other materials can be used instead of reed stems? What are the advantages of a thatched roof compared with a tiled roof? What are the disadvantages?

9 Here is a map quiz. An atlas will help you to answer these questions about the map below, which shows the part of the United States known as the Corn Belt (see page 15).
(a) Five of the States are shown by their initial letters. What are the names of these states?
(b) What is the name of the lake which projects down into the Corn Belt?
(c) What is the name of the city on the shores of the lake?
(d) Is the Corn Belt mainly flat or is it mountainous?
(e) What is the name of the large river which flows across the Corn Belt? Where does it flow into the sea?

# Savanna grasslands

Dry season on the savannas

The grasslands which grow in tropical areas are called savannas. Savannas are found in those parts of the world, near the Equator, where it is hot all the year round. They grow in places where there is hot, wet weather. This is followed by dry seasons when little or no rain falls. Most trees cannot grow in these places. Those few that do grow have roots which go deep down into the soil. With these long roots the trees can obtain water from deep underground in the dry season.

The savanna grasslands are found in South America, Africa and northern Australia. They are flat plains and plateaux. During the wet season, the grasses grow tall and green. Any trees on the savanna bear leaves and flowers then. In the dry season little or no rain falls. The trees lose their leaves. The grass turns brown. The grass plants go into a resting stage. They do not grow again until rain falls.

Often fire sweeps across the savanna during the dry season. It may burn down the trees and shrubs and produce ash. This helps the grasses to grow better when the rains return.

The distribution of savanna grasslands (shown in green)

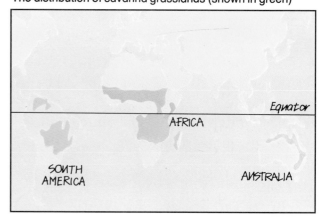

Equator
AFRICA
SOUTH AMERICA
AUSTRALIA

Bush fire

# The savannas of Africa

The savannas of Africa are the homes of some of the world's most interesting animals. The large animals are often called 'game animals' or 'big game'.

Over the grassy areas, huge herds of antelopes, buffaloes and zebras roam. Elephants, giraffes and rhinoceroses feed among the thorny bushes and scattered trees. All of these animals feed on plants. They are herbivores. Some of the herbivores migrate huge distances in the dry season to the parts of the savanna where the grass is still growing. The herbivores provide food for carnivores such as lions, cheetahs, leopards, hyenas and hunting dogs.

Large areas of the African savannas have been set aside as national parks and game reserves. Here the animals can live in peace. Tourists are able to visit the parks or reserves to see the big game animals. Although hunting is not allowed, poaching is a problem in many game parks and reserves. This is the illegal hunting of animals. Some poachers are hungry people from outside the park or reserve who are looking for meat. Some poachers want only the hides, tusks or horns of the animals to sell to tourists for high prices.

*Top*: Wildebeest crossing the River Mara
*Right*: Elephant poachers

27

# People in the African savannas

Masai women, and (*inset*) a Masai tribesman draws a bow to puncture the steer's vein for blood

Some of the people of the African savannas are herdsmen. The Masai people live on the borders of Kenya and Tanzania. The Masai build their huts near a river or water-hole. From their huts, the young Masai men and their herds of cattle move slowly across the surrounding grassland. When all the grass has been eaten, they build huts in a new area. The Masai feed largely on cow's blood mixed with milk. They sometimes obtain potatoes and flour and beans from other tribes in exchange for meat and hides.

Some other savanna tribes are farmers. During the dry season they clear a small area of savanna. They do this by burning the plants. The ashes help to fertilize the soil a little. When the rainy season comes the people plant their crops. After a few years the plant foods in the soil have been used up or washed away. The people move to a new area and make a new clearing. This type of farming is often called shifting cultivation.

There are many large modern farms in the African savanna. These can exist because they have enough water. This water is often used to irrigate the crops. Sugar cane, maize, coffee, groundnuts, cotton, tobacco and pineapples grow well on the savanna soils if they receive enough water.

Spraying young coffee plants

# The Australian savanna

Much of northern Australia consists of savanna grassland. The region has the typical savanna climate. The summers are hot and wet. They are followed by long hot, dry winters. Much of the grassland is used for rearing beef cattle.

Although the grass grows lush and green in the wet summers, it is not very good as food for the cattle. In the winter, the grass is brown and dry and there is little water for the cattle to drink. As a result, it takes more than 20 hectares of land to feed each animal. The cattle farms are usually called cattle stations. They are very large. Each may cover several thousands of square kilometres.

For much of the year the cattle roam far and wide. In the autumn they are rounded up. The calves are branded, while some of the older animals are sent off to market. The cattle stations are often a long way from the nearest town. In the past the cattle had to be walked to market.

A cattle station

Sometimes this took many weeks. The animals became thin and weak on the journey and many died. Nowadays most cattle are taken to market in 'road trains'. Each road train can quickly carry up to 90 cattle to market.

Cattle at a water-hole near Broome

A typical road train

# The South African veld

The grasslands of South Africa are usually called the veld. This is a Dutch word for field. Parts of the veld are more than 1200 metres above sea level. These are known as the high veld. In many ways the veld is like the savanna grasslands further north. But the veld is usually drier. And because it is nearer to the sea the climate is not so hot. Occasionally snow falls on the high veld.

Originally the veld was covered with wild grasses. There were also areas of scrub and trees. The animals of the veld were similar to those of the savanna. Giraffes, elephants, lions, cheetahs, wildebeest and other kinds of antelopes lived there. Huge herds of antelopes called springboks roamed across southern Africa. They were named because of their habit of leaping into the air when frightened or playing. Some of the herds contained more than a million animals.

Farm landscape at Grey's Pass, West Cape, South Africa

Today much of the veld has been destroyed and springboks are now quite rare. Some of the veld is used for grazing sheep. Mealies (maize) is grown on parts of the high veld. Many of the remaining parts of the veld, and the animals on them, are protected as game reserves and national parks.

Springboks at Etosha, Namibia

# The Canterbury Plains of New Zealand

Sheep farming is one of the main industries of New Zealand. Many of these sheep are reared on the Canterbury Plains on the South Island of New Zealand. The Canterbury Plains are the largest area of flat land in New Zealand. They lie at the foot of the Southern Alps.

The Alps cut off the Plains from many of the rain-bearing winds. Because of this the Canterbury Plains have only a moderate amount of rain. The sunny climate and occasional rain means that grass grows well all the year round.

Originally the Plains were covered with wild tussock grasses. Gradually the early settlers changed the Plains into good farming country. Today most of the wild grassland has been replaced by cultivated grasses and crops. Even where the wild grassland has not been ploughed up, heavy

The Canterbury Plains of New Zealand

grazing by rabbits and sheep has changed it. The farmers grow crops of wheat, oats and potatoes on the Canterbury Plains. Grass and turnips are grown as food for the sheep. After the sheep have been killed their carcases are frozen. They are exported to other countries in refrigerated ships.

Sorting sheep into pens near Waiau

Frozen lamb being loaded at the docks

# Prairies

North American Indians hunting bison

The prairies of the United States and Canada were named by French explorers in the 16th century. The word 'prairie' means meadow or grassland. Trees are rare on the prairies except along river banks and around the farm buildings.

The prairies lie roughly half way between the Equator and the polar regions. They receive more rainfall than many other grasslands. Most of this rain falls in the spring and early summer and helps the grass grow quickly.

Great herds of pronghorn antelopes and bison ("buffalo") used to feed on the North American prairies. Bison are wild cattle. They were hunted by the American Indians. From the bison the Indians obtained meat, and also hides for

The prairies of Alberta, Canada

their clothes and tents. In the 19th century when white people tried to cross and settle in the prairies, they often fought with the Indians. The European settlers used the prairies first as cattle or sheep ranches. They killed off most of the bison, originally for food, later just to get rid of them. Eventually the prairies were used to grow wheat.

# Wheat growing on the Canadian prairies

Harvesting on the prairie

Canada is one of the world's largest wheat producers. The wheat is grown on the Canadian prairies. As we have seen, once these prairies were wild grasslands. Today most of them have been ploughed up and planted with crops.

The prairie farms are very large and a long way apart. Many of the farm buildings have steep roofs so that the snow slides off them in winter. Often they have tall hedges or rows of trees around them to give shelter from the wind.

The wheatfields are enormous. They stretch as far as the eye can see. Because of this, they are ideal for the use of large machines. Ploughing, sowing and harvesting are all done by machines. On the prairies the growing season is short because of the bitterly cold winters. As a result, varieties of wheat which ripen quickly have to be planted.

Harvesting begins in August in the southern prairies after the hot, dry, sunny summer. The ripe grain is taken by lorries to storage towers called elevators. These are built by the side of the railway tracks. Later much of the grain is taken by rail to the ports. There it is stored in more elevators until ships carry the grain to all parts of the world.

Notice the silos, pond, and windbreaks on this farm in Saskatchewan

Grain elevators by the railway track

# The dust bowl soil erosion

An abandoned farm in the 'dust bowl', Oklahoma, May 1937

As we saw on page 4, grass helps to stop the soil from being washed or blown away. Where the soil is dry and dusty, and there is no covering of grass, the soil is easily blown away. The finest particles of soil blow away first. The larger pieces of soil which are left do not grow plants well.

During the 1920s large areas of the natural prairie grassland were ploughed up. They were sown with wheat, another kind of grass. For several years the harvests were good. But in the 1930s there were several very dry years. The wheat crop died and the soil lay bare and unprotected. The dry weather turned the soil to dust. And the strong winds blew up huge dust storms. Houses, farms and fences were buried by the drifting soil. The area became known as the 'dust bowl'. The soil was ruined and so were thousands of farmers.

The Soviet Union also ploughed up dry areas of the steppes after the Second World War. Here, too, serious soil erosion occurred. This kind of soil erosion can be prevented by careful farming.

34

# The Pampas

The pampas are the vast grassy plains of South America. They are among the flattest grasslands in the world. Most of the pampas are in Argentina, but parts are in Uruguay and Brazil. Near the sea they receive abundant rain. They are covered by tall tussocky grasses. But the western pampas are blocked from the rain-bearing winds by the Andes mountains. As a result they are dry and mostly barren.

Originally the pampas were almost empty of people and large animals. Only a few tribes of American Indians wandered across the pampas. It was they who gave the grasslands their name. The rhea, the maned wolf and the armadillo fed on the pampas. But then in the 16th century invaders arrived from Spain. The Spaniards brought cattle and horses with them. Some of these were turned loose by the early settlers. Soon vast herds of cattle and horses were running wild over the grassy plains.

At first the cattle were simply hunted for their meat and hides. Then it was realised that the pampas were ideal for rearing cattle. These animals could be left out in the open to feed all the year round. It was not possible to sell the cattle overseas because they were thin and weak by the time they arrived. But the discovery of how to preserve food by canning meant that meat from the

Inset: pampas grass

Brazil

Argentina

P A M P A S

Uruguay

SOUTH AMERICA

The pampas of South America (shown in yellow)

pampas could be exported. Later, the invention of refrigerated ships made it even easier to export good quality meat.

Young armadillo feeding

The giant anteater wanders across the pampas

# The changing pampas

During the second half of the 19th century, large farms were built upon the pampas. The fields were divided up by barbed wire fences. These made it easier to control the breeding and grazing of the cattle. Pedigree bulls were taken from Europe to the pampas to improve the quality of the local cattle.

Before long much of the pampas had changed. Most of the tall 'pampas grass' had gone. In its place were fields of alfalfa or lucerne which were grown to feed the cattle. Deep wells and wind pumps dotted the landscape. These provided the water needed for the crops and cattle.

The damper eastern parts of the pampas are now farmed. There are huge fields of wheat, oats, rye, maize and flax. Railways built across the flat pampas have encouraged the

New industrial buildings compete with farmland at Santa Fe, Argentina

growth of towns and villages. The railways carry the cattle and grain to the ports. Once cowboys called gauchos rounded up the wild horses and cattle on the pampas. Now their descendants work on the big farms, helping to raise cattle and cereal crops. Today, rounding up the animals is done using motor vehicles.

Rounding up cattle in Argentina

# The steppes

The steppes are the largest grasslands in temperate parts of the world. They are named after a Russian word for 'plain'. These dry, treeless, grassy plains are far away from the sea. They stretch from Romania, through southern Russia into China.

In summer, the steppes are hot, dry and dusty. In winter the ground is frozen hard and covered with snow. On most of the steppes the grass is fairly short and dry. When the spring thaw comes, the grasses and other plants grow rapidly. Then until the middle of May, the steppes are bright with flowers.

Once bison and vast herds of wild horses and saiga antelope roamed the steppes. Now all but the saiga antelope have gone. The wolves which preyed on the herds of grazing

The steppes are shown in green

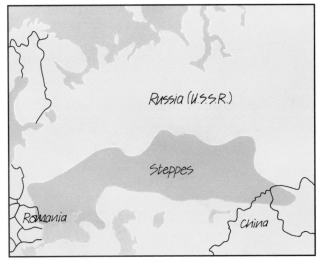

animals have become rare. Today sheep and horses graze parts of the steppes. In fact it was on the steppes that people first domesticated horses. The grass is too poor and the winters too cold for cattle. However, most of the original steppe grassland has now gone. Like the prairies, the steppes have largely been changed into vast wheat and barley fields.

**Siberian steppes in summer**

**A winter encampment on the Mongolian steppes**

**Przewalski's horses on the Mongolian steppes**

# The saiga antelope

Male saiga

Female saiga

Saiga antelopes live on the steppes of Russia. The saiga is about the size of a sheep. The male has horns. Once thousands of saiga roamed the steppes, eating the poor, dry grass. Because people believed that the horns of the saiga could be used as a medicine, many animals were killed during the 19th century. Only the saigas' horns were taken. Their flesh was left to rot.

By 1918 the saiga were almost extinct. Only a few hundred remained. In 1919 the government banned the hunting of the saiga. At the same time, scientists began to study the life history of these antelopes. Soon the numbers of saiga increased. By 1970, 2 million of them were roaming the steppes. There were too many of them for the amount of grass on the steppes.

Eventually, someone realised that the saiga could be useful. Now each year about 250,000 saiga are killed for their meat. This tastes like mutton. The saiga skins are also used as leather. This meat and leather comes from a part of the steppes which is too poor to feed sheep or cattle. The soil is too dry and the weather too cold for crops to grow. This was one of the first examples of modern people treating wild animals like a crop.

# Vanishing grasslands

Many kinds of natural grassland are becoming rare. In the cooler parts of the world the grasslands were easy to plough and improve. They have now largely been taken over for growing cereals. Some are used for grazing cattle and sheep. There is little left of the natural pampas, prairies, or steppes.

Many of the tropical grasslands too have been changed by people. The damper areas of the savannas and veld have been taken over for crops. The drier areas are used for rearing cattle. Those parts of the savanna which remain are under threat. The number of people in the world is increasing rapidly. This means that more food is needed. Most of this has to come from new farmland.

In just over 100 years, many of the huge herds of animals which roamed the savannas have been destroyed. We shall have to take great care to preserve the natural grasslands and wild grazing animals that are left. We have already seen how useful the wild saiga antelopes have become to the Russians. In the future it may be possible to use other grassland animals in a similar way.

Grassland that has been taken over for growing crops

Thomson's gazelles may be farmed in the future

# Do you remember?

1 What are savannas?

2 How do fires on the savanna help the grass to grow better?

3 What is poaching?

4 Why do people poach in the African game parks and reserves?

5 Where do the Masai people live?

6 What are the main foods of the Masai?

7 What is meant by shifting cultivation?

8 What are the cattle farms on the Australian savanna called?

9 How are animals taken to market from the farms on the Australian savanna?

10 Where do the veld grasslands occur?

11 How did springboks obtain their name?

12 What is the veld used for nowadays?

13 What kind of farming is carried out on the Canterbury Plains of New Zealand?

14 What shelters the Canterbury Plains from many of the rain-bearing winds?

15 Where are the prairies?

16 What does the word 'prairies' mean?

17 Why were the bison ('buffalo') on the prairies hunted?

18 What crop is grown on the Canadian prairies today?

19 Why do the prairie farm buildings have steep roofs?

20 What are elevators?

21 What caused the soil erosion in the area known as the 'dust bowl'?

22 Where are the pampas?

23 Which early settlers released cattle and horses on the pampas?

24 What discoveries and inventions made it easier to export meat from the pampas?

25 What was done to improve the quality of the pampas cattle?

26 What are gauchos?

27 Where are the steppes?

28 Why are cattle not kept on the steppes?

29 What are the steppes used for today?

30 What do the Russian people obtain from the saiga antelope?

# Things to do

**1 Ornamental grasses** Some seed merchants and garden centres sell packets of seeds of ornamental grasses. These are the seeds of grasses from all over the world.

Sow some of these seeds in the open garden in summer, or in pots of moist compost indoors. Watch them grow. The flowers look very attractive cut and put in vases with other garden flowers. They can also be dried and used as decorations in the winter.

Some seed mixtures sold as food for canaries, budgerigars and other tame birds also contain the seeds of interesting grasses. Sow a little of this bird seed and see what grows from it.

Ornamental grass seed

**2 The prairies** Draw or paint a large picture of the prairies as they might have been before they were given over to growing cereal crops. In your picture you might show herds of bison ('buffalo'), Red Indians and a wagon train of settlers or their camp.

**3 A model prairie farm** Make a model of a prairie farm. Use cardboard, clay, plasticine, drinking straws, twigs and any other materials you can get. Paint your model.

A prairie farm in British Columbia

**4 A breakfast cereal survey** Carry out a survey amongst your friends at school. Find out which is their favourite breakfast cereal. Which is the least popular cereal? Why? Make a block graph or histogram of your findings.

Find out which cereal grains the different breakfast cereals are made from.

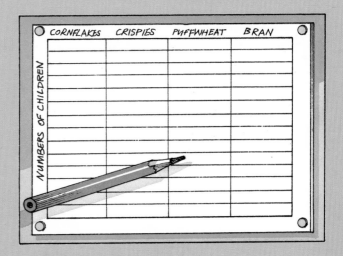

**5 Make a picture** Draw or paint a large picture of workers on an Argentinian cattle station rounding up the cattle ready for market (see page 36).

**6 Where will grasses grow?** We read on page 4, that grasses will grow almost everywhere.

Try to grow grass seeds on different moist surfaces. As well as soil or compost you could try moist sand, ashes, felt, blotting paper, newspaper, cloth, sawdust, cardboard, orange peel, a piece of sponge, pieces of broken brick. How many grass seeds did you sow? How many of them grew? On which of these materials does the grass seed grow best?

**7 Make a water wheel** Often in the past the power of running water was used to turn the machinery which ground cereals to make flour. There are still many of these watermills remaining, although few of them still work.

The water of the river or stream turned a water wheel. You can make a model water wheel to see how it works.

Cut the metal foil from a yoghurt pot or cream carton. Make a small neat hole at the centre. Cut 8 slits at equal distances apart around the edge. Twist the pieces between the cuts to make paddles. Push a knitting needle through the hole in the middle of the wheel.

Hold the water wheel under a running tap. Does the wheel go round? This is how a water wheel works.

Collect pictures of real watermills. Make a wallchart with your pictures.

**8 Making a toy windmill** Another way in which cereal grains used to be ground into flour was by using a windmill. A windmill uses the power of moving air, or wind, to turn its machinery.

Cut a square of thin card or stiff paper so that it has sides about 15 centimetres long. Make four cuts in the card or paper as shown in the picture. Make five small holes as shown.

Take a piece of thin wire. Bend over the end and then thread a small bead on to the wire. Twist the corners of the card or paper over the centre so that the five holes are lined up. Thread the wire through the holes and then through another small bead.

Twist the end of the wire around a thin stick. See that the windmill will turn freely. Hold the windmill up to face a breeze, or run with it.

Where were windmills built to ensure their sails turned rapidly? What could be done with some windmills if the wind was blowing from the wrong direction?

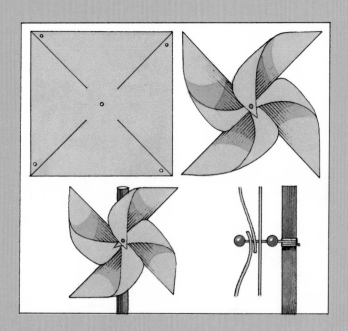

**9  Quick bread**  On page 23, we saw
how to make unleavened bread. Normally
bread is made to rise, so that it is light and
spongy, with the help of yeast. This process
is rather slow. Here is a way of making
bread quickly, using self-raising flour
instead of yeast to make the bread rise.

*You will need:*  450g self-raising flour
                1 heaped teaspoon salt
                3 tablespoons vegetable
                   oil
                Milk
                A grown-up to help you

*What you do:*

1  Place the flour in a basin and stir in the
   salt.
2  Put the vegetable oil in a measuring jug
   and add enough milk to make it up to
   300 millilitres.
3  Whisk the oil and milk together and pour
   them on to the flour and salt. Stir gently
   until all the flour is moistened. Press the
   mixture lightly with your fingers and
   then turn it onto a floured surface.
   Knead the dough with your fingertips
   until it just holds together in a single
   lump.

4  Place the dough, smooth side uppermost,
   on a greased baking tray. Flatten the
   dough with your knuckles until it forms a
   thick round. Use a knife to cut a deep
   cross in the top of the dough.
5  Brush the surface of the dough with milk.
   Then bake it in a hot oven (425°F, 210°C,
   Gas Mark 7) for 20 to 25 minutes until
   the loaf is golden brown and firm to the
   touch.

   How does the appearance and taste of
   this bread differ from that of the
   unleavened bread?

OXFAM

274 Banbury Road
Oxford OX2 7DZ

Save the Children

Mary Datchelor House,
17 Grove Lane,
Camberwell,
London SE5 8RD

*Areas of recent famine*

Enough food for everyone?

**10 Famine** Many countries in the African savanna lands and in the rice-growing areas of the world suffer from famine or starvation from time to time.

Make a list of the countries where people are suffering from famine now or have suffered from it in recent years. Make a sketch map of the world to show where these countries are. Do these countries have hot or cold climates, are they wet or dry for much of the year? Are there any other reasons why people in a country may suddenly not have enough food?

What are the better-off countries doing to help those countries where there is famine? Could they do more to help? If so, say how or what you think they could do.

**11 Make a picture** Use different cereal grains and breakfast cereals to make a picture. Choose a fairly simple shape such as a bird or a house. Glue your cereals on to the card or wood to make the shape you want.

**12 Life among the grass stems**
Pretend you are only the size of an ant. You live amongst the grasses on a lawn or grassy field. Write a story describing a day in your life.

**13  Sheep and cattle**  Collect pictures of the different breeds of sheep and cattle. Make a scrapbook or wallchart of your pictures. Say whether each breed of cattle is kept mainly for milk or beef. Say whether the sheep are kept mainly for meat or wool.

**14  Animals of the African savanna**  Collect pictures of the animals of the African savanna. Make a wallchart or a book of them. Write a sentence or two about each picture, saying which of the animals are herbivores and which are carnivores.

**15  Hairy egg-people**  Here is a different way to grow grass seed.

Collect some empty egg-shells and wash them in soapy water. Rinse the egg-shells in cold tap water and allow them to dry.

Make a smooth lump of plasticine and gently push a clean egg-shell into it. Fill the egg-shell with wet cotton wool. On the outside of the egg-shell draw a face. You can use poster paint, enamel paint or a felt-tipped pen for this. Sprinkle some grass seeds on top of the wet cotton wool.

Stand some of the egg-men on a warm, sunny windowsill. Put others in a dark cupboard. Water the egg-men every day and wait for their "hair" to grow. How long does this take and what colour is the hair?

Make a similar egg-man and stand it in the refrigerator. Does the hair grow?

Take another egg-man but this time leave the cotton wool dry. Does the grass seed grow?

# Things to find out

1  Find out how the wheat-growing season in Canada differs from that in, say, Britain or France. Are the same kinds of wheat grown in all three countries?

2  Find out how rice can be grown on hill slopes.

3  Why is most of the work of planting, weeding and harvesting rice in the paddy fields done by hand?

4  Draw a map of Kenya. Mark on it as many of the game parks and game reserves as you can find out about.

5  The Masai people (page 28), do not have a settled home. They lead a wandering or nomadic life. Find out about other people in other parts of the world who are nomads.

6  Find out why the elephant and the rhinoceros are favourite targets for poachers in the African savanna game parks and reserves.

7  On the large grasslands such as the pampas and the Australian savannas, cattle are kept for beef rather than for their milk. Why is this?

8  The wild game animals on the African savanna can provide much more meat than cows kept on the same area of savanna. Find out why this is. Would it be better to domesticate antelopes and other game animals on the savanna rather than cows? Why?

9  Milk is a food which comes indirectly from grass. Find out the name of the dairy which delivers the milk you drink at home or school. Do you know where the dairy is? Where does it get its milk from? What is done to the milk before it is put into bottles or cartons? How is the milk taken to the dairy?

10  In our recipe on page 43, we made the bread rise using self-raising flour. Usually bread is made to rise using yeast. Find out more about yeast: where it comes from, what else it is used for, and how it is used.

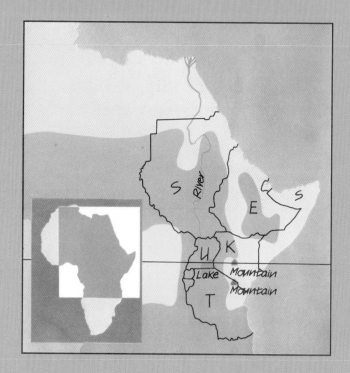

11  Here is a map quiz. An atlas will help you to answer these questions about the map above which shows part of the savanna lands in East Africa.
(a)  Six of the countries are shown by their first letters. What are their names?
(b)  What are the names of the two mountains shown on the map? Which is the higher of the two?
(c)  What is the name of the large lake? Roughly how long is it?
(d)  What is the name of the large river which flows across the area shown on the map? Where does it flow into the sea?
(e)  What is the name of the line drawn right across the map?

# Glossary

Here are the meanings of some words which you might have met for the first time in this book.

*Carnivore:* a hunting flesh-eating animal.

*Cereals:* certain members of the grass family which are grown for food.

*Elevator:* usually a lift, but also a tower on the prairies used for storing grain.

*Equator:* the imaginary line around the centre of the Earth.

*Fertilizer:* a chemical substance containing plant foods.

*Game animals:* wild animals that are hunted for sport or food.

*Game reserve:* an area of land where game animals are protected; a nature reserve.

*Grain:* the seeds of a cereal.

*Hay:* dried grass which is used as a food for farm animals, especially in winter.

*Herbivore:* an animal which feeds on plants.

*Irrigation:* ways of bringing water to fields to help crops to grow.

*National Park:* a large area of land over which special care is taken to see that the beautiful scenery is not spoilt and where the plants and animals are protected.

*Nomads:* groups of people who have no settled home, but who wander from place to place.

*Paddy:* another name for rice when the grains are still in their husks.

*Paddy fields:* flooded rice fields.

*Pampas:* the name for the great fertile grassy plain of South America, most of it in Argentina. The pampas are one of the great beef-rearing and wheat-growing areas of the world.

*Pasture:* grassland used for feeding farm animals.

*Plantation:* a large farm or estate which specializes in crops to be sold, such as rubber, sugar cane, cacao, coffee, etc.

*Poaching:* the illegal hunting of animals for their meat, hides, horns, tusks or other parts.

*Polar regions:* the very cold parts of the Earth around the North and South Poles.

*Prairie:* the wide grassy plains of Canada and North America, now used mainly for growing cereals.

*Savanna:* the tropical grasslands of Africa, Australia and South America.

*Scavenger:* an animal which feeds on dead animals which it has not itself hunted or killed.

*Silage:* grass or other green crops pressed and stored away from the air for later use as food for cattle.

*Soil erosion:* the gradual wearing away of soil by wind or water.

*Tropics:* the land or region around the Equator.

*Tundra:* the cold, treeless plains in the northern and Arctic regions.

*Veld:* grassland with few trees in Southern Africa.

*Weed:* a wild plant growing where it is not wanted.

# Index